EARTH AND SPACE

Written by
Martin Lunn

Illustrated by
Brigid Collins, Line & Line, The Maltings Partnership, and Steve Noon

Edited by
Penelope Lyons

Designed by
Charlotte Crace

Picture research by
Helen Taylor

CONTENTS

The Sun

It is not surprising that many ancient civilisations regarded the Sun as a god. It is the largest and brightest object in the sky and its appearance in the morning chases away the darkness of night. The ancient Egyptians called their sun god Ra. The ancient Greeks and Romans believed in the same sun god – he was called Apollo.

The Sun is a huge ball of burning gases. It appears so large in the sky because it is very much bigger than the Earth and it is close by. The Sun totally rules the Solar System. If the Sun was not there, we would not be here. There would be no Earth. The Sun provides light and heat without which life cannot exist.

Did you know?

The Sun is about 150 million km from the Earth.

Its diameter is about 1.4 million km.

The temperature in its centre is about 14 million °C.

The Egyptian sun god, Ra

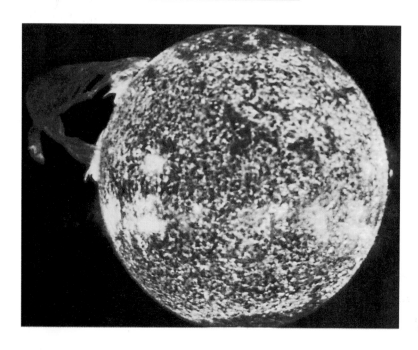

This 'flame' is called a solar flare

Suns and stars

The Sun is a star and is very similar to many of the other stars we can see in the night sky. All stars are suns, so we could say that our Sun is our local star. If we were to journey out to the stars and look back towards the Sun, it would look no different from the other tiny dots we see in the sky.

Some stars are younger than our Sun, some older. At present our Sun is about 'middle-aged'. It will eventually use up all its gases and will 'die'. Luckily for us it still has about another 5000 million years' worth of gases left!

The Sun has a surface temperature of about 6000°C. The hottest stars have surface temperatures of over 50 000°C, the coolest are about 2600°C.

Safety first

You must be very wary of the Sun. It is very hot and bright. You must never look directly at the Sun through a telescope or binoculars. You will blind yourself if you do. Even staring at the Sun can cause blindness.

People sometimes say 'red sky at night, sailors' delight' because they think a dramatic red sunset means tomorrow will be a nice day

Sun observations

The Sun rises in the east and sets in the west, so during the day the Sun appears to move across the sky. This movement is not along a straight flat line; the Sun appears over the horizon and then climbs up the sky during the morning and drops down towards the horizon again in the afternoon.

Shadows through the day

Shadows can tell us about the Sun's position. Shadows are long when the Sun is low in the sky; they are short when the Sun is high in the sky. Shadows are shorter at midday than in the morning or evening.

A shadow stick is a piece of wood either permanently placed in the ground or attached to a base as part of a moveable system. The Sun must be shining brightly enough to cast a sharp shadow. If the shadow stick is kept in one place, the length and position its shadow can be measured at different times of the day.

Sundials are really shadow sticks for telling the time

The shadow moves round the stick during the day. The position and length of the shadow changes because the Earth is spinning and the position of the Sun appears to have moved.

Understanding shadows enabled people to make sundials. The shadow cast by the shadow stick or gnomon falls onto a scale so you can tell the time.

Shadows through the year

The height of the Sun in the sky changes through the year as well as through the day. This is because the Earth moves round the Sun. In the United Kingdom, shadows are shortest in June and longest in December because the Sun is highest in the sky in June and at its lowest in December. In Australia, the shortest shadows occur in December and the longest in June.

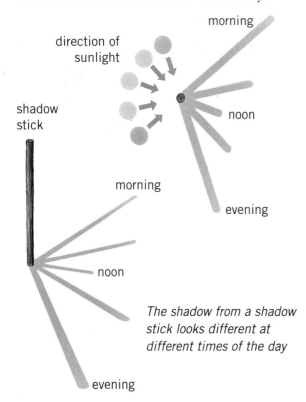

The shadow from a shadow stick looks different at different times of the day

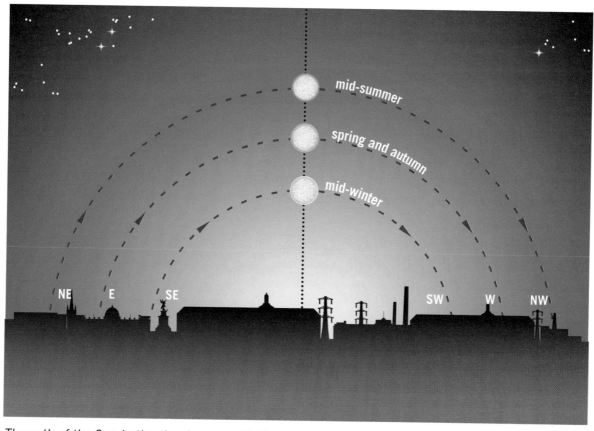

The path of the Sun in the sky changes with the time of year (you are facing south)

This gnomon (shadow stick) was used by astronauts on the Moon to measure the position of the Sun

Measuring the Earth

Greek scientists of 2500 years ago observed the Sun and made calculations about the size of the Earth. Their estimates in 240 BC were better than those used by Christopher Columbus in 1492. Columbus was looking for India when he found America. The Earth was bigger than he thought!

Day and night

Our planet is called the Earth, it is a sphere or ball of rock that travels around the Sun in its own particular path. This path is called the Earth's orbit. At the same time, the Earth is spinning. It spins all the time, around an imaginary line passing through its centre from the North Pole to the South Pole. This imaginary line is called the Earth's axis.

The Sun is shining all the time, even when we can't see it. It is the spinning of the Earth that causes day and night. It is daytime when your side of the Earth is facing towards the Sun. It is night when your side of the Earth is facing away from the Sun.

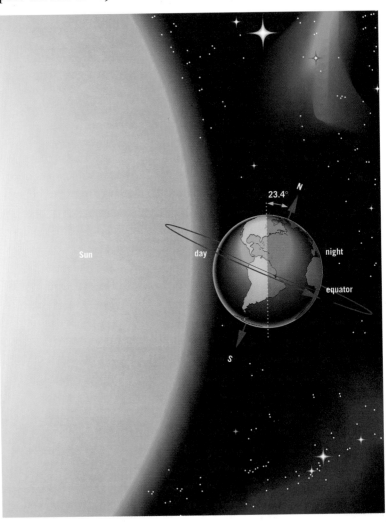

The side of the Earth facing the Sun has daytime

Turning night into day
Throughout history daytime has been seen as a period of safety and night as a time of danger.

In the very earliest times, people used to strike two sharp stones together to make sparks, then they could light fires. By Roman times, people could light their homes using lamps with vegetable oil or animal fat. During the Middle Ages, people knew how to make and use candles to provide light. A hundred years ago, people used gas lamps. Today we use electric light bulbs.

Daytime and the 24-hour day

The Earth takes 24 hours to spin round once. Astronomers call this period of time a day. This can be confusing because in that 24-hour day we have the time we usually call a day (when it is light) and the time we usually call night (when it is dark). An ordinary day plus a night equals an 'astronomical day'.

Although we now know that day and night are caused by the Earth spinning, there have been many different beliefs as to why there is day and night.

The ancient Greeks and Romans believed the sun god Apollo drove his chariot across the sky in the day and rested at night

Day and night on other planets

All the planets in the Solar System have day and night because they all spin. Some of them take longer than the Earth to spin round once and some take less time. The ones that take longer have daylight and night for a longer time than the Earth. The ones that take less time have daylight and night for less time.

Day and night on Saturn

The seasons

The ancient Egyptians thought that the Earth took one year (12 months or 365 days) to orbit once round the Sun. We now know that an Earth year is a bit longer – 365.25 days. Each year has four seasons, spring, summer, autumn and winter.

The Earth is divided into two halves, the northern and southern hemispheres, by an imaginary line round the middle called the equator. The Earth does not 'sit' upright in space, it is tilted over to one side by 23.4°. As the Earth moves round the Sun, this tilt never changes. First one hemisphere is tilted towards the Sun then, when the Earth is on the other side of the Sun, the other hemisphere is tilted towards it.

Summer and winter

When the hemisphere you live in is tilted towards the Sun, you can see the Sun for longer, there is more light and heat and it is summer. Six months later, your hemisphere will be tilting away from the Sun so you will receive less light and heat. This is when you have winter. In between are spring and autumn.

The Earth spins once in 24 hours whether it is winter or summer. In summer when your hemisphere is tilted towards the Sun, it appears to be higher in the sky and the days are very long. In the winter, the Sun appears to be much lower in the sky and the days are shorter.

All the planets go round the Sun and they all have four seasons.

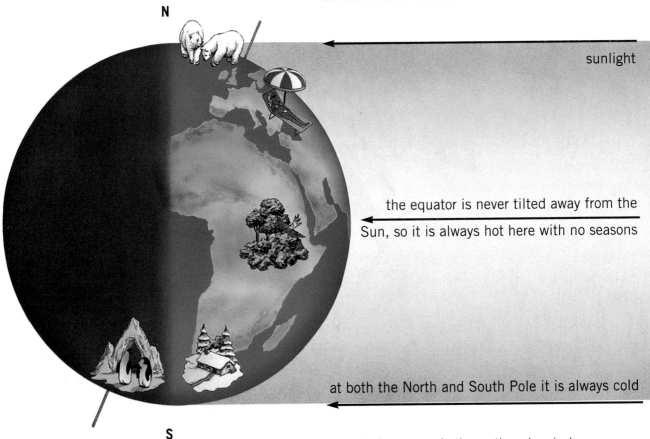

N

sunlight

the equator is never tilted away from the Sun, so it is always hot here with no seasons

at both the North and South Pole it is always cold

S

When it is summer in the northern hemisphere, it is winter in the southern hemisphere

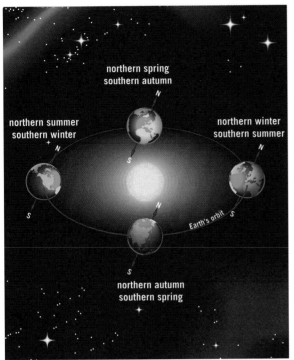

What season it is depends on whether or not your hemisphere is tilted towards the Sun

North and south

The length of daytime is also affected by how far north or south of the equator you live. In June, the United Kingdom has about 15 hours of daylight. Further north, you can still see the Sun in the time we usually call night. This is called the midnight sun. At the North Pole during summer it is always light because the Sun never sets. At the same time of year, New Zealand has only about 8 hours of daylight and at the South Pole the Sun can't be seen at all because the South Pole is tilted too far away from the Sun.

Six months later, the southern hemisphere has the longer periods of daylight during its summer, while we have winter. In our winter the Sun can't be seen at the North Pole, but is always visible at the South Pole where it is summer.

Cycle of seasons				
Northern hemisphere	spring	summer	autumn	winter
Southern hemisphere	autumn	winter	spring	summer

In the northern hemisphere, you have to travel a long way north in summer to see the midnight sun

The Moon

The full moon

The crater furthest away is one of the biggest on the Moon and is called Copernicus

The Moon is our closest neighbour in space. It is a ball of rock that orbits the Earth. It is Earth's only natural satellite.

After the Sun, the Moon appears as the biggest and brightest object in the night sky. However, appearances can be deceptive. The Moon seems to be so large because it is very close to us in space terms. In fact, it is only about one-quarter the size of the Earth.

The Moon has no light of its own. We see it because it reflects sunlight. Moonlight is reflected sunlight.

The Moon's surface is covered by dents or holes called craters. These are mainly caused by lumps of rock crashing onto the Moon's surface.

The Moon is the only other place in the Universe visited by human beings. In 1969 Neil Armstrong and Buzz Aldrin became the first people to walk on the Moon. So far 12 men have visited the Moon but no women.

Atmosphere

The Earth is surrounded by layers of air. These layers are called Earth's atmosphere. The Moon does not have an atmosphere. It is airless and there is no wind or rain. If you go for a walk on a beach, you leave footprints in the sand. Later, the tide comes in and washes them away. Or, if the sand is dry, the wind might blow sand into your footprints and cover them up. But if you visited the Moon, you would still be able to see the astronauts' footprints from 1969. The footprints made by the astronauts and the craters made by rocks will remain on the Moon's surface for ever because there is no atmosphere on the Moon.

There is nothing to remove these footprints from the Moon's surface

Moon myths

Ancient civilisations considered the Moon as a god and today there are still many interesting and unusual stories, jokes and legends about the Moon. For instance, that it is made of 'green cheese', that you might go mad when there is a full moon, and that certain unfortunate people turn into werewolves on the night of the full moon!

Werewolves only appear when there is a full moon!

Moonwatch

If you have a pair of binoculars, you can use them to look at the Moon. When you look at the Moon, you can see some grey areas. These are mountains and craters which make the 'Man in the Moon's face'. (Do not try this at the full moon as it can be painfully bright.)

Have you noticed that sometimes you can see the Moon during the day as well as at night?

The Moon is about to be 'eclipsed' by this hot air balloon!

Why does the Moon appear to change shape?

The Moon takes 29.5 days to orbit round the Earth. It takes the same time to rotate once on its own axis. In this time its appearance changes. This is because light from the Sun always illuminates half of the Moon, but sometimes the illuminated half of the Moon is facing away from us.

When the Moon is between the Sun and Earth, its dark side is turned towards us. When this happens we cannot see the Moon and we say there is a new moon. But when the Earth is between the Sun and the Moon, we see the lighted side of the Moon. When we can see *all* of the lighted side, we say there is a full moon. In between times, we see part of the lighted side; sometimes we can see a lot of it, sometimes only a little. The different shapes or phases of the Moon have names such as a crescent moon and a gibbous moon.

We never see the far side of the Moon. This is because it takes the Moon the same time to rotate on its axis as it takes it to orbit the Earth.

Sun's rays

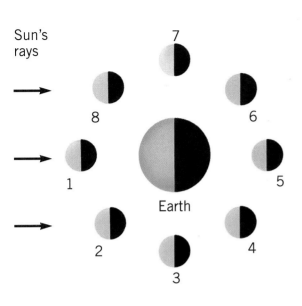

Earth

Phase of the Moon		Time of day when seen
1 new moon		(not seen)
2 waxing crescent		most of the day and early evening
3 first quarter		afternoon and first half of the night
4 waxing gibbous		evening and most of the night
5 full moon		all night
6 waning gibbous		most of the night and early morning
7 last quarter		second half of the night and morning
8 waning crescent		before sunrise and most of the day
9 new moon		(not seen)

Why we don't see the far side of the Moon

Try sitting on a stool (you are the Earth) and ask a friend to walk round you as the Moon. Your friend must keep looking at you all the time so you always see your friend's face but never see the back of your friend's head. In the time that your friend has walked round you, he or she has also turned round themself once. Try this slowly and you will see how your friend turns round to always face you. This is how the Moon behaves.

Eclipses

Eclipses happen because the Earth and the Moon cast shadows. Try taking a ball outside on a sunny day and throwing it against a wall. As the ball moves, so does its shadow. The shadow the ball casts on the ground or wall is like the shadow the Moon casts on the Earth.

When the Moon passes between the Earth and the Sun so that the Sun, Moon and Earth form a straight line in that order, the Moon casts a shadow on the Earth. This is called an eclipse of the Sun or a solar eclipse. If you are standing in the Moon's shadow, you can see the Sun gradually disappear then reappear as the Moon passes in front of it.

In a solar eclipse the Moon hides the Sun

When the Earth passes between the Sun and the Moon so that the Sun, Earth and Moon are in a straight line in that order, the Earth casts a shadow on the Moon. This is called an eclipse of the Moon or a lunar eclipse. The Moon is not hidden during a lunar eclipse, it turns a coppery red as the Earth's shadow falls on it.

Predicting eclipses

We do not have eclipses of the Sun and Moon every month. The Moon orbits the Earth at an angle so they do not often form a straight line with the Sun. However, because we know about the movements of the Earth and the Moon, we can predict when eclipses are going to occur.

In a lunar eclipse the Moon turns a coppery red

Viewing a solar eclipse

Even during an eclipse of the Sun you must not look directly at the Sun through binoculars or a telescope. To be safe, you can stand under a leafy tree and look at the ground. As the Moon passes in front of the Sun, you will see lots of images of the dark circle of the Moon with light from the Sun behind it.

A dangerous dragon
In ancient China it was believed that a solar eclipse was caused by an unfriendly dragon (the Moon) passing in front of the Sun and trying to eat it. To prevent this and to scare the dragon away, as much noise as possible was made by banging drums, beating gongs and shouting. It always worked!

A safe way to view a solar eclipse

Discovering the planets

In ancient times astronomers noticed that there were five stars that wandered about the sky. These are planets. The word 'planet' comes from the Greek verb 'to wander'.

The planets Uranus, Neptune and Pluto were not discovered until the telescope was invented. The ancient astronomers did not know about them because they are too far away to be seen without optical aids.

Like the Moon, the planets do not produce light. They reflect some of the sunlight that reaches them. This is why they shine.

The first five planets

Like the ancient astronomers, you can see the bright planets without the help of binoculars or a telescope. They are not all visible at the same time because they have different orbits. Newspapers produce charts and information about the night sky at the end of each month. This will tell you which planets you can see.

▶

Mercury is difficult to see because it appears close to the Sun. It looks like a bright pinkish star. Look for it when the Sun has just set.

◀

Venus is the brightest of the planets. It is sometimes called the morning star or the evening star because it is in the eastern sky before sunrise or the western sky after sunset. But remember, it is a planet not a star.

Mars is very red. Occasionally (when it comes very close to Earth) it becomes the second brightest planet.

Jupiter is usually the second brightest planet. If you have a pair of binoculars, you can sometimes see up to four of Jupiter's largest moons.

How long is a year?

Some planets have years that are longer than ours, and some are shorter. Each planet travels round the Sun in its own orbit. The length of time a planet takes to go round the Sun once is called its year. The planets that are closer to the Sun have much shorter orbits than the planets that are further away. This means they have shorter years.

Saturn is the most distant planet you can see without a telescope. It is not so bright as the others. It looks like a dirty yellow star.

Facts and figures about the planets

Name	Distance from Sun (km)	Diameter (km)	Time taken to spin once (Earth time)	Time taken to go round Sun once (Earth time)	Surface temperature (°C)	Number of moons or satellites
Mercury	58 million	4878	58d 15h 36m	88d	−180 to +430	0
Venus	108 million	12 102	243d 0h 14m	225d	465	0
Earth	150 million	12 756	23h 56m 04s	365.25d	15	1
Mars	228 million	6786	24h 37m 48s	687.98d	−133 to +22	2
Jupiter	778 million	142 984	9h 55m 30s	11.8yr	−150	16
Saturn	1427 million	120 536	10h 39m 22s	29.4yr	−180	18
Uranus	2871 million	51 118	17h 14m	84yr	−210	15
Neptune	4497 million	49 528	16h 7m	164.7yr	−210	8
Pluto	5914 million	2284	6d 9h 18m	248.5yr	−220	1
(yr = years, d = days, h = hours, m = minutes, s = seconds)						

Planets and gods

The five 'wandering stars' were named after gods: Mercury was the messenger, Venus was the goddess of love, Mars was the god of war, Jupiter was the ruler of the gods, and Saturn was the father of Jupiter.

When the remaining three planets were discovered they were also named after gods to fit in with the others. Neptune was the god of the sea. Uranus was the oldest of all the gods and was the father of Saturn. Pluto was the god of the underworld. The Earth is the odd one out, it is the only planet not named after a god. This is because it is the one we live on.

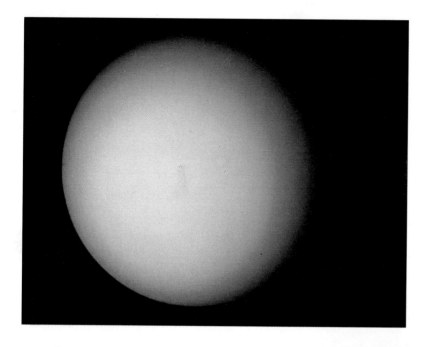

Uranus was discovered by William Herschel in 1781, 170 years after the astronomical telescope was invented.

After another 65 years, in 1846, Johann Galle and Heinrich D'Arrest discovered Neptune.

▼

Cylde Tombaugh discovered Pluto in 1930. He made the discovery as a result of mathematical calculations. This photograph was taken by the Hubble Space Telescope; it is the only picture we have of Pluto.

▼

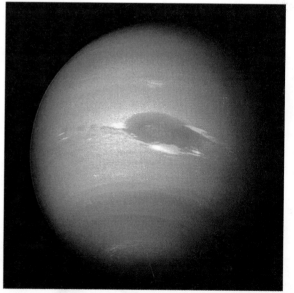

The Solar System

The Earth is one of nine planets orbiting the star called the Sun. Some of the planets have satellites or moons orbiting them. The Sun, the planets and their moons, the asteroids, comets and other small pieces of dust, make up what we call the Solar System.

The inner planets

The four inner planets are Mercury, Venus, Earth and Mars. They are all small and rocky bodies. They have very few moons or none at all.

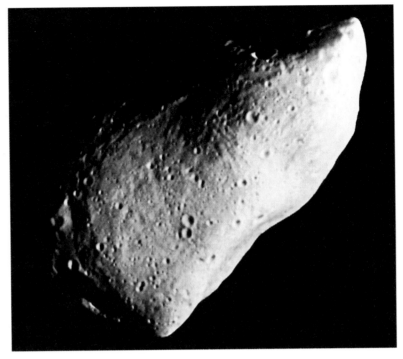

This asteroid is called Gaspra

Asteroids

Beyond Mars is the asteroid belt. It is made up of thousands of lumps of rock that orbit the Sun together. Asteroids are as old as the rest of the Solar System.

The outer planets

There are four large outer planets. They are often called gas giants. Jupiter, Saturn, Uranus and Neptune are all large spheres of gas. They are much colder than the inner planets. They all have large families of satellites or moons.

In 1930, Pluto was found. It is the outermost planet and is not a gas giant. It is a small world which appears to be made of ice and rock. It is the only planet which no space probe has visited.

Comets

Comets are minor members of the Solar System. They are lumps of rock, dust trapped in ice. They look a bit like snowballs that are full of dirt and grit and are often called 'dirty snowballs'.

Halley's Comet is the most famous and brightest comet

Meteors and meteorites

Meteors are tiny pieces of dust which orbit the Sun. Sometimes they rush into the Earth's atmosphere and burn up. When this happens you can see streaks of light in the night sky. Rather confusingly these are often called shooting stars. At certain times of the year the Earth passes through dense clouds of dust and we experience meteor showers.

Meteorites are rocks large enough not to burn up in the Earth's atmosphere. The last big one was about 20 metres across and entered the atmosphere in 1908. It destroyed over 2000 square kilometres of forest when it exploded 5 miles above northern Siberia.

The five largest asteroids

Name	Diameter (km)
Ceres	1003
Pallas	608
Vesta	538
Hygeia	450
Euphrosyne	370

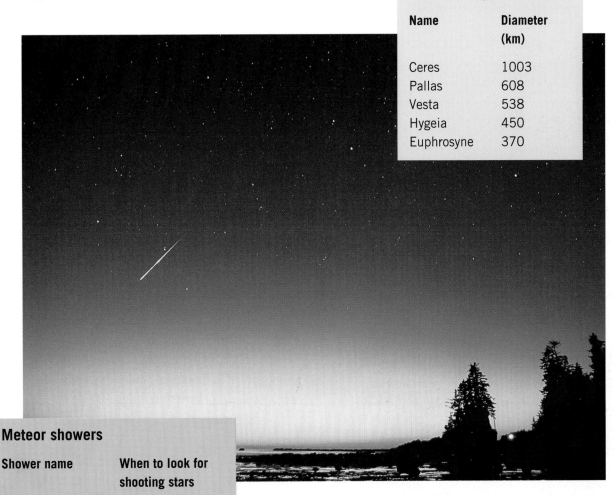

This meteor was photographed in Canada at dusk

Meteor showers

Shower name	When to look for shooting stars
Quadrantids	January 3–4
Lyrids	April 22
Eta Aquarids	May 4–5
Delta Aquarids	July 28–31
Perseids	August 12–13
Orionids	October 21
Leonids	November 17
Geminids	December 13–14

The stars

Take a look at the night sky on a good clear night. The stars you see are suns just like ours only much further away, that is why they appear so small.

On a clear night from anywhere on Earth it is possible to see about 2500 stars without the help of a pair of binoculars or a telescope.

Stars are formed in clouds of gas that are sometimes called stellar nurseries. The Orion Nebula is one of these. The star map on page 31 will help you find it.

Stars go through different stages and then die. Some finish their lives very quietly; others explode into a very bright star called a supernova before fading.

Star types

There are lots of different star types and sizes. Small stars are called dwarfs and the large ones are called giants. The very large ones are called supergiants. If you look very carefully at the stars, you can see that they are different colours.

The star colours tell astronomers how hot a star is. The stars which appear blue or white in the sky are much hotter than those which appear red. Our Sun is a yellow dwarf.

Galaxies

Galaxies are groups of millions of stars. Some galaxies are called spiral galaxies because of their shape, but when a spiral galaxy is viewed from the side, it looks rather like two fried eggs back to back.

blue supergiant

orange giant

red dwarf

yellow dwarf

red giant	orange giant	yellow giant	blue giant
red dwarf	orange dwarf	yellow dwarf	white dwarf
red supergiant	orange supergiant	yellow supergiant	blue supergiant

These twelve types of star are all different sizes

Seeing stars

In the daytime we can only see one star, the Sun. The other stars are there – we just can't see them because the light from the Sun is so bright that it overwhelms the fainter light from the stars.

Stars don't really twinkle, but they do appear to. This is due to the Earth's atmosphere. The starlight has to pass through the atmosphere before reaching the ground. The layers of air that make up our atmosphere are always moving, so the light from a star seems to move about slightly. This movement is the twinkle.

Navigating

The stars can be used as a compass. If you can find the Great Bear or Plough in the sky, you can then find the Pole Star. This is very useful because the Pole Star tells you where north is. The map on page 29 will help you find the Plough and the Pole Star.

The Orion Nebula is an area where stars are formed

M74(NGC 628) is a spiral galaxy about 30 million light years away

Solar System and star trivia

Earth
Our home is often called the Blue Planet. This is because it has a lot of water in the oceans and looks blue from space.

Mercury
Only one spacecraft, Mariner 10, has visited Mercury. Before that, the best drawings of the planet were made in France during the 1930s. Some of the areas on Mercury were given very strange names, such as 'The Wilderness of Hermes the Thrice Greatest'!

Sun

Venus
This is the only planet named after a female god or goddess. It is not the closest planet to the Sun, but it is the hottest. It has a thick atmosphere that keeps heat in. The Venusian day is longer than the Venusian year. Venus and Uranus are the only planets to spin in a clockwise direction.

Mars
Mars has ice caps that seem to be frozen carbon dioxide. It also has the largest known volcano in the Solar System, Olympus Mons, which is 600 km across and 24 km high.

Jupiter
The largest planet in the Solar System has the shortest day because it spins very fast. A day on Jupiter lasts 9 hours and 55 minutes of Earth time. Ganymede is Jupiter's largest moon, it is also the largest moon in the Solar System and is bigger than Mercury or Pluto. Like Saturn, Jupiter has rings but we can't see them from Earth.

The stars
The largest stars are called supergiants, the smallest are dwarfs. A supergiant is almost as big as the Solar System. A white dwarf is about the same size as Pluto.

Neutron stars are the result of a supernova explosion. They have diameters of about 30 km.

The Sun
The Sun sometimes has black dots on its surface. These are sun spots and are cooler areas on the Sun. Many sun spots are larger than the Earth!

Saturn

Photographs of Saturn taken from Earth show Saturn's rings. The rings look solid but they are made of millions of very tiny particles that orbit round Saturn.

Neptune

This is the windiest planet in the Solar System. It has wind speeds reaching 2000 km/h. Neptune has rings but we can't see them from Earth.

Uranus

This planet is tipped over on its side at an angle of 98° and, like Venus, it spins clockwise. Uranus also has rings but we can't see them from Earth.

The Solar System

The Solar System is mainly empty space.

Pluto

Pluto is the smallest planet in the Solar System and is farthest from the Sun.

Constellations and mythology

In the past, astronomers drew patterns among the stars and gave the patterns the names of heroes and villains. The stars in each pattern or constellation have no special connection. They just happen to be in the same part of the sky.

The ancient Greek astronomers knew 48 constellations. Today we know 88. The new ones are mainly in the southern hemisphere. The Greeks didn't know about them because they didn't travel there.

Orion the Hunter

Orion is a splendid constellation. Orion was a hunter who boasted that he could kill any living creature. Unfortunately, he forgot about the scorpion which stung him on the ankle and killed him.

Orion and the Scorpion are both in the sky, but the gods decided that they should never meet again. In the northern hemisphere, Orion is in the winter sky, and the Scorpion is in the summer sky.

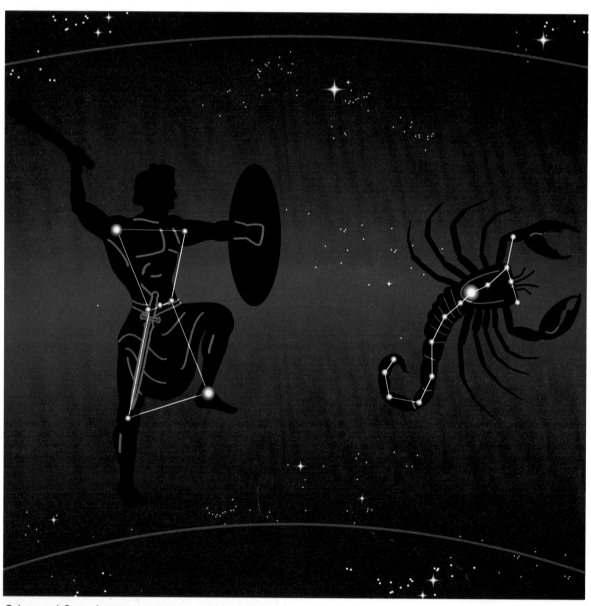

Orion and Scorpio never meet

Perseus rescues Andromeda from the sea monster

The Great Bear

The Great Bear or Plough can only be seen in the northern hemisphere. Its story is about Callisto and Arcas.

Callisto was very beautiful and had a son called Arcas. Juno, the queen of the gods, was very jealous of Callisto's beauty and turned her into a bear. Many years later, Arcas was out hunting and was about to shoot a bear, because he didn't know it was really Callisto. Jupiter, the king of the gods, saved Callisto by turning Arcas into a bear too. He then caught both animals by the tails, and swung them up into the sky. That is why the Great Bear (Callisto) and the Little Bear (Arcas) have long tails.

Naming the new constellations

The southern constellations were mainly discovered by European explorers in the seventeenth and eighteenth centuries. These constellations were often named after new scientific instruments such as the microscope and the telescope.

Cassiopea

All the characters in Cassiopea's story can be found in the sky. Queen Cassiopea was very boastful and one day she upset Neptune, the king of the sea. He sent a sea monster called Cetus to attack Cassiopea's land. The only way Cassiopea could prevent disaster was to chain her daughter Andromeda to a rock for Cetus to eat. At the last minute, Perseus arrived on the winged horse, Pegasus. Perseus was carrying the Medusa's head and when Cetus looked at it, the monster was turned to stone. In this way Perseus rescued Andromeda whom he then married.

The Great Bear and the Little Bear

Signposts in the sky: the Great Bear

The Great Bear or Plough is visible all year. The dashed lines on the star map are the lines you imagine in the sky to help you find other stars.

Finding Polaris

First find the Great Bear. Now you can find Polaris, the Pole Star. Imagine a line drawn from Merak through Dubhe and then carrying on till you reach a bright star on its own. This is Polaris.

Finding Cassiopea

Now imagine a line from Alioth through Polaris. Carry on for some distance on the far side of Polaris. You will reach five bright stars arranged in a rough 'W'. This is the constellation Cassiopeia.

Finding Capella

If you imagine a line from Megrez through Dubhe and keep going, you will eventually come to Capella. On winter evenings Capella is high up and may pass directly overhead. On summer evenings Capella is at its lowest and almost reaches the horizon. Capella is yellowish and is accompanied by a small triangle of stars close by.

Finding Vega

You can find Vega by imagining a line from Phad, to Megrez and then curving out and carrying on till you reach a star that looks blue. This is Vega.

Finding Castor and Pollux

The twins Castor and Pollux can be found by imagining of a line from Megrez through Merak and carrying on until you find the two stars close together. Castor and Pollux can only be seen in winter.

Finding Arcturus and Spica

Arcturus is orange and about as bright as Capella or Vega. To find Arcturus, you must imagine a line from Mizar through Alkaid and then curve downwards. If you continue along the curve past Arcturus you will find another bright star, Spica. Arcturus and Spica can be seen in spring and early summer.

Names and names

The stars were named by many peoples, including the Chinese, Indians and Aztecs. Different civilisations have different names for the same group. We still use the names given by the peoples of the Mediterranean. But sometimes we translate them into English, and sometimes we have other names as well. For example, we often call Ursa Major, the Great Bear. We also call it the Plough. In America, it is called the Big Dipper.

CASSIOPEA

VEGA

POLARIS

CAPELLA

MIZAR

DUBHE

ALKAID

MEGREZ

ALIOTH

MERAK

CASTOR

PHAD

POLLUX

ARCTURUS

REGULUS

SPICA

Signposts in the sky: Orion

Orion can only be seen in winter and all its chief stars are brilliant. The dashed lines on the star map are the lines you imagine in the sky to help you find other stars.

Orion's Belt

The three stars in the middle of the constellation are called Orion's Belt. They are Mintaka, Alnilam and Alnitak. Just below Orion's Belt is the stellar nursery, the Orion Nebula. This is a fuzzy patch in the sky. If you look closely, you will see a star surrounded by a fuzzy area.

Finding Sirius

From Alnitak imagine a line pointing to the left and downwards and you come to Sirius. Sirius is also known as the Dog Star. Sirius is the most brilliant star in the sky but it is less bright than Venus or Jupiter.

Finding Aldebaran

Imagine a line from Mintaka going upwards and to the right. Carry on past Bellatrix and you will come to Aldebaran.

Finding Procyon and Alphard

If you imagine a line drawn from Bellatrix through Betelgeux then carrying on and curving slightly downwards, you will come to Procyon. This is sometimes called the Little Dog Star. You can continue the line to the left and downwards and you will discover Alphard.

Finding Castor and Pollux

Orion can also help you find the twins, Castor and Pollux. Imagine a line going from Rigel to Betelgeux and then carry on. You will come to the two stars close together that you will recognise as Castor and Pollux. This will help you link the signposts together.

Finding Capella

Capella can also be found from both signposts. Imagine a line going from Saiph to Alnitak and then passing between Betelgeux and Bellatrix and carrying on for quite a long way. Eventually you will come to Capella.

Can you find your way back to the Great Bear?

Did you know?
The largest constellation is Hydra and the smallest is the Southern Cross. The constellation with the greatest number of bright stars is Orion and the constellation with the fewest visible stars is Mensa.

POLLUX

CASTOR

CAPELLA

ALDEBARAN

PROCYON

BETELGEUX

BELLATRIX

ALPHARD

MINTAKA

ALNITAK ALNILAM

ORION NEBULA

SAIPH

RIGEL

SIRIUS

Do the constellations move?

The stars seem to swing round the Pole Star during the night. This is because the Earth is spinning. The constellations that are not close to the Pole Star appear to move into and out of the night sky during the year. This is because the Earth is orbiting the Sun.

As the Earth moves round the Sun, we see different parts of space and different constellations, but we can always see the Pole Star and the stars close to it. Just as when you look at opposite walls of your bedroom, you can always look up and see the ceiling.

Circumpolar stars

These are the stars that are visible all year round. In the United Kingdom, the Plough is circumpolar, but we never see the Southern Cross. In the southern hemisphere the Plough can never be seen but the Southern Cross is circumpolar.

Star gazer brolly

To make a star gazer brolly (SGB) you will need an old umbrella and some sticky luminous stars.

Open the umbrella. Think of the centre (where the handle joins the canopy) as the Pole Star. Now stick some of the luminous stars inside the brolly in their correct constellation positions. You can use the maps on pages 29 and 31 to help you.

Hold the SGB above your head and slowly turn it. You can see how the constellations seem to rotate round the handle, or Pole Star. This is how the stars seem to move at night. You can test this for yourself.

Choose a non-cloudy evening in autumn or winter when it gets dark early. Go out and look at the position of the stars at 5.00 p.m. and line up your SGB with them. Fix your SGB securely, then look at the stars again at 8.00 p.m. You will see how much the stars moved. This is because the Earth is spinning.

A star gazer brolly

The ten brightest stars	
Star	**Constellation**
Sirius	Canis Major
Canopus	Carina
Alpha Centauri	Centaurus
Arcturus	Bootes
Vega	Lyra
Capella	Auriga
Rigel	Orion
Procyon	Canis Minor
Achernar	Eridanus
Betelgeux	Orion

Star trails are seen in 'time-lapse' photographs because the stars seem to move round the Pole Star during the night

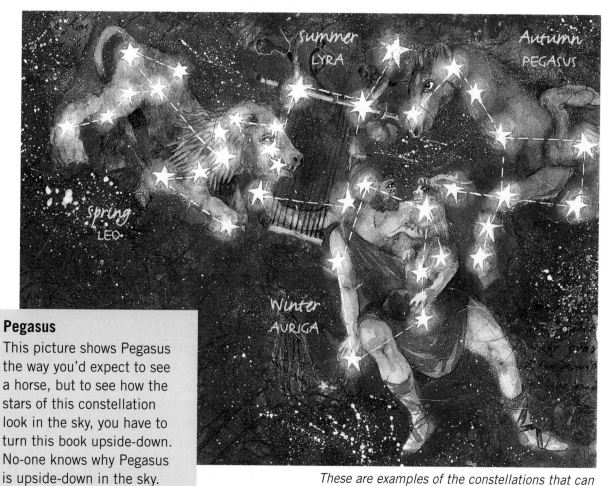

Pegasus
This picture shows Pegasus the way you'd expect to see a horse, but to see how the stars of this constellation look in the sky, you have to turn this book upside-down. No-one knows why Pegasus is upside-down in the sky.

These are examples of the constellations that can be seen in each season

Searching the sky

Astronomy is the science which deals with the stars and planets. Astronomers study this topic.

Observatories and telescopes

Today, the astronomer's main tool is the telescope. Astronomical telescopes were invented in the early seventeenth century. Observatories are research stations containing telescopes and much other equipment.

There are two main types of light telescope: refractors and reflectors. Refracting telescopes use lenses to collect light, but reflecting telescopes use mirrors to do the same thing. All professional telescopes are now computer-controlled.

On Earth, observatories have to be built well away from cities and bright lights because of light pollution. They are usually found on the tops of mountains where the air is very dark and clear.

The largest telescope on Earth

Living conditions
Because observatories are often in lonely places, astronomers have to make sure they have enough food and drink and somewhere to sleep.

The world's largest mirror telescope is the Keck telescope in Hawaii. It has a mirror 10 metres across. The Hubble Space Telescope is not the largest, but it is in the best place – in space orbiting the Earth and well away from bright lights.

Light pollution

The full moon provides useful light during the night. Astronomers often get annoyed with the presence of the full moon because the light drowns the faint light of the stars, making them difficult to see. This is a natural form of light pollution.

Artificial light is a much bigger problem for astronomers. Cities and towns keep getting bigger. There are industrial centres and shopping centres with more and more street lights. People who live in large cities or near large shopping centres find it very difficult to see any stars at all.

From space you can see where all the major centres of industry and population are by the amount of light. It looks very impressive.

Providing adequate light at night is very important so people can see where they are going and be safe. However, if street lights were designed to direct all their light downwards, then less energy would be used. This would be good for everyone because roads and paths would be brightly lit and the night sky would remain dark.

The Hubble Space Telescope being launched from the space shuttle

See for yourself
Next time you are in a car or on a train at night, look at the glow of the lights above the towns and cities. This is light pollution.

City lights in the British Isles seen from space at night

Space travel

Space travel to explore the Solar System and beyond, means very long journeys. It is important to learn how to live and work in space for very long periods of time; at least a year and probably longer.

Since the 1970s, the Americans and Russians have launched space stations which orbit the Earth and allow astronauts to work in space. A new multi-national space station is likely to be put in space next century.

Casual clothes are not very different in space ...

What happens to water

It is important not to have water droplets floating around a spacecraft because without gravity the droplets will eventually run together and form one big 'floating lake'.

Inside the spacecraft

Working and living conditions in space are very different from those on the Earth. In space there is no air or gravity. Astronauts can take their own air with them, but they cannot take gravity. In space, astronauts float around their spacecraft rather than walk around it.

Scientists have tried to make life as normal as possible for the astronauts. Inside the spacecraft, astronauts wear their normal space clothes. They eat their food from a tray using a knife and fork, but drinking has to be done from a special bottle with a stem.

Going to the lavatory is very different in space. Waste material in the lavatory doesn't fall down the bowl. It has to be sucked away by air. And astronauts have to wear a seat belt to stop them from floating away.

... but playing an instrument is

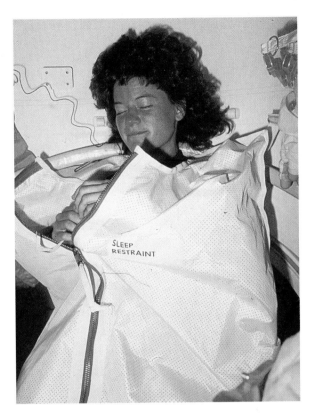

Astronauts sleep zipped into sleeping bags that are attached to the floor or wall. If they are not fastened down, the astronauts could drift across the cabin and possibly injure themselves.

Out in space

Outside the spacecraft, astronauts must wear spacesuits. Spacesuits protect the astronaut from harmful cosmic rays; they also have air tanks attached to them for the astronaut to breathe.

To work outside, astronauts may use a manned manoeuvring unit (MMU) – a sort of flying armchair that allows them to move away from the spacecraft. The MMU is moved by firing little jets of gas out of the backpack, like mini rockets.

You have to sleep strapped down...

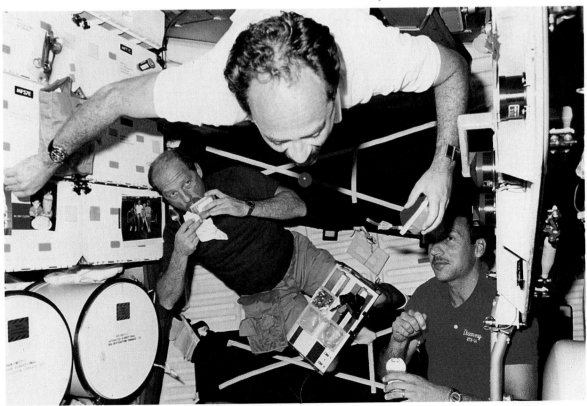

... but drink might float free

Spin-offs from space

Space technology affects our life today in all sorts of ways.

In the home

One of the first materials to be developed for use in space was Teflon. Nothing sticks to it and it is now used for non-stick frying pans.

ISBN 0-563-39785-3

9 780563 397854 >

Tools such as cordless drills were originally developed because astronauts working in space required tools that had their own power sources. After all, they couldn't plug a tool into an electric socket in space.

Solid lubricants were also developed for use in space. The most famous one is WD–40. Many people spray this onto their car engines to help them start.

In the supermarket

Sandwiches are often sold in a clear plastic triangular wrapper. This packaging was originally developed to keep food fresh for the astronauts in the 1960s.

Plastic thinly coated with metal was also developed in the 1960s as insulation material. Today, it is found as wrappers for crisps and sweets.

There are millions of parts in a space shuttle. Every part is numbered electronically so that a computer can read its number. Supermarkets and shops use this technology to keep a check on their stock. You know the system used – it is the bar code you see on anything you buy from a supermarket.

Clothes

Mylar is another insulating material used in space exploration. Now you can find it in sleeping bags and outdoor jackets. Ambulance paramedics use Mylar blankets to keep injured people warm until they get to hospital.

To help astronauts train for weightless conditions, training shoes with pockets of air in their soles were developed in the mid-1980s. Many sports shoes now have this type of sole.

In the workplace

A lap-top computer is a small portable computer. These were developed in the 1980s because astronauts are very busy in space and do not have the time to keep going back to the main computer. They need a small computer that can be carried around.

Extra info
Velcro was invented long before the space programme but it was used in the 1960s in the Gemini and Apollo spacecraft. It lined the walls of the spacecraft so astronauts could stick things onto it to prevent them floating away.

Is anybody there?

**We do not know if there is any other life in the Universe.
We are certain there is no other life in the Solar System.
The planets closer than Earth to the Sun are too hot;
those further away are too cold.**

Mars is the planet that most resembles Earth. In 1976, two spacecraft landed on Mars to look for life. They did not find any but they did send back photographs from the Martian surface. Today, it looks dry and bleak and is cold. We don't know if microbe life forms existed there when a thick atmosphere and water were present. We will have to wait until manned missions visit Mars to find the answer.

The Universe is huge

The Solar System is a tiny region of space ruled by the Sun. Our Sun is one of about 100 000 million stars that make up our galaxy, the Milky Way. It is a spiral galaxy. There are millions of galaxies in the Universe. The Milky Way is neither the largest nor the smallest.

In the southern hemisphere it is possible to see some nearby galaxies. They are called the Large Magellanic Cloud and the Small Magellanic Cloud. They were first observed by Ferdinand Magellan who sailed round the world in 1519–21.

On a very clear night it is possible to see the constellation Andromeda and a small 'fuzzy' area in space. This is the Andromeda galaxy; it is 2.2 million light years away. It is the most distant object you can see without a telescope or binoculars.

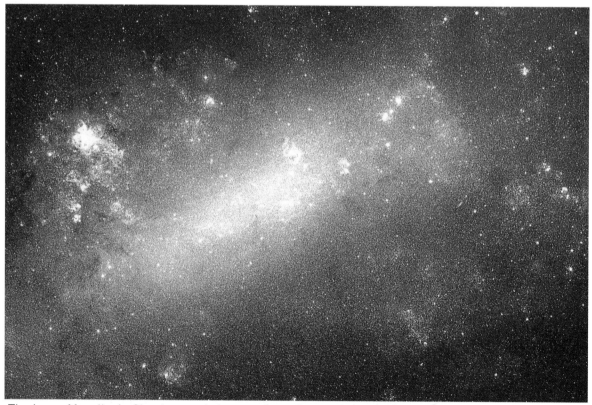

The Large Magellanic Cloud can only be seen from the southern hemisphere

Looking for life

Project SETI (Search for Extraterrestrial Intelligence) uses powerful radio telescopes to look for messages from space, but so far, there have not been any.

Space probes have been sent into deep space beyond the Solar System. These spacecraft carry messages that might be found by another intelligent civilisation.

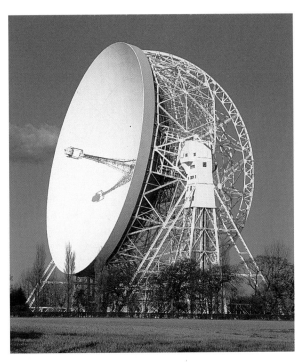

The radio telescope at Jodrell Bank looks for messages from space

We imagine all kinds of other life forms

The ten closest stars to us	
Star	**Distance in light years**
Proxima Centauri	4.2
Alpha Centauri A	4.3
Alpha Centauri B	4.3
Barnard's Star	6.0
Wolf 359	7.7
BD + 36° 2147	8.2
Sirius A	8.6
Sirius B	8.6
UV Ceti A	8.9
UV Ceti B	8.9

The speed of light

Light travels extremely fast but the Universe is so big that it can take years for light from a star to reach us. A light year is the distance travelled by light in one year; it is about 9.5 million million km or 6 million million miles.

Astronomy logbook

BC

2500
Astronomical
records begin

585
An eclipse of
the Sun is first
predicted

350
Aristotle
suggests the
Earth is round

240
The size of
the Earth is
estimated

140
The first star
catalogue is
drawn

1957
Sputnik 1 is the first
artificial satellite

1959
The first photograph
of the far side of the
Moon is taken

1961
Yuri Gagarin is the
first man in space

1962
Mariner 2 to Venus is
the first successful
probe to another planet

1963
Valentina Tereshkova is
the first woman in space

1969
Neil Armstrong and
Buzz Aldrin are the first
men on the Moon

1976
Viking 1 and 2
land on Mars

AD

813
The Baghdad
School of
Astronomy is
founded

1543
Copernicus says
the Earth is not
the centre of the
Universe

1610
Galileo uses the
first astronomical
telescope

1675
The Royal Greenwich
Observatory is founded

1781
Uranus is
discovered

1801
The first
asteroid is
discovered

1840
The first photograph
of the Moon is taken

1846
Neptune is
discovered

1930
Pluto is
discovered

1981
The space shuttle
is tested

1990
The Hubble Space
Telescope is launched

1991
The Keck telescope
is built – it is the
largest in the world

1994
Comet Shoemaker/Levy 9
smashes into Jupiter

The Zodiac

The ancient astronomers noticed that the five bright planets they could see were only ever seen in the same part of the sky as certain constellations. These constellations are in a band across the sky. The path that the Sun appears to travel during the day is in the same band. This band is called the Zodiac.

Zodiac means 'circle of animals'. It is the name for the collection of constellations where the planets are seen during an Earth year. They are mostly named after animals. The Zodiac was discovered by the ancient astronomers and is now used by astrologers. Astronomers study the stars as a science. Astrologers think that your life is affected by the positions of the planets and the stars and the Sun when you were born.

There are thirteen constellations in the Zodiac. Twelve of them are the ones whose names appear as the signs of the zodiac or star signs in newspapers and magazines. The extra one is called Ophiuchus, the Serpent Bearer. It is found between Scorpius and Sagittarius.

The planets Mercury and Pluto have orbits that sometimes take them out of the main band of the Zodiac. These planets can sometimes be seen in other constellations.

Do you know all the constellations in the zodiac?

ARIES

AQUARIUS

PISCES

TAURUS

GEMINI

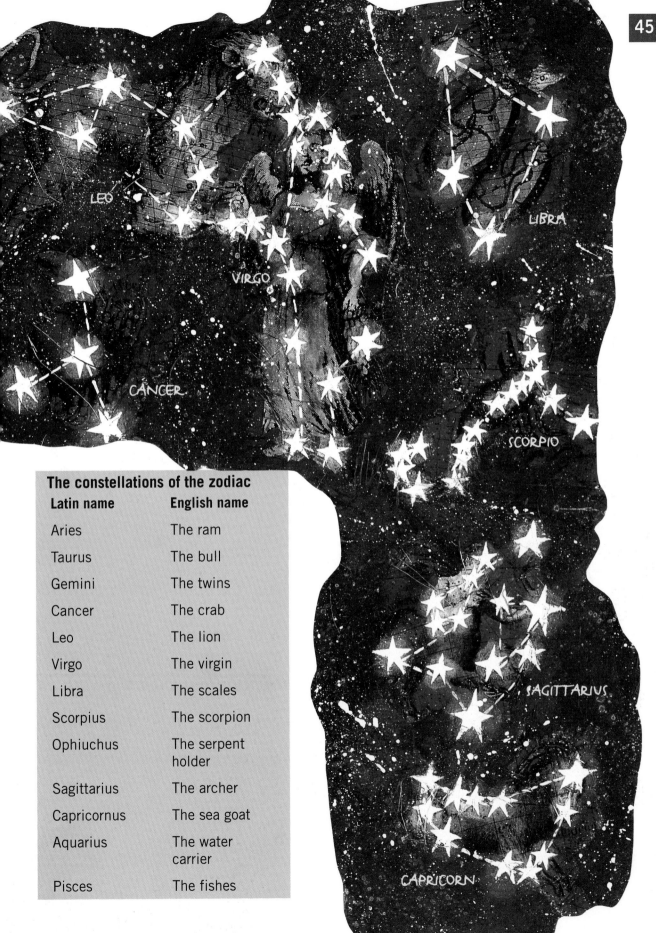

LEO

VIRGO

LIBRA

CANCER

SCORPIO

SAGITTARIUS

CAPRICORN

The constellations of the zodiac

Latin name	English name
Aries	The ram
Taurus	The bull
Gemini	The twins
Cancer	The crab
Leo	The lion
Virgo	The virgin
Libra	The scales
Scorpius	The scorpion
Ophiuchus	The serpent holder
Sagittarius	The archer
Capricornus	The sea goat
Aquarius	The water carrier
Pisces	The fishes

Glossary

Astronomy The science dealing with stars and planets

Atmosphere The layers of gas surrounding a planet

Circumpolar star A star which circles the Pole Star

Comet A 'dirty snowball' made of rock, and dust in ice

Constellation A pattern or group of stars

Day The period taken by a planet to spin once on its axis

Eclipse (lunar) The time when the Moon passes into the shadow cast by the Earth

Eclipse (solar) The time when the Moon passes in front of the Sun

Galaxies Star systems with thousands of millions of stars

Gnomon The pointer of a sundial

Halley's Comet The only bright comet whose return can be predicted; it returns to the Sun every 76 years

Light year The distance travelled by light in one year; it is equal to about 9.5 million million kilometres or 6 million million miles

Meteor A tiny particle in space that burns up when it dashes into the Earth's atmosphere and produces a shooting star

Meteorite A rock large enough to survive burning up in the atmosphere and that may produce a crater when it falls to Earth

Midnight sun The Sun seen above the horizon at midnight during summer within the Arctic or Antarctic Circle

Observatory An astronomical research station

Orbit The path of a planet, satellite or asteroid

Phases of the Moon The apparent changes in shape of the Moon

Planets The most important members of the Solar System apart from the Sun

Plough The nickname for Ursa Major, the Great Bear

Polaris The Pole Star

Proxima Centauri The nearest star beyond the Sun

Seasons Differences in weather and temperature due to the tilt of the Earth as it orbits the Sun

Shooting star The popular name for a meteor

Solar System The system made up of the Sun, planets, satellites, comets, asteroids, meteorites, dust and gas

Star A gaseous body that produces its own light (also called a sun)

Sundial An instrument used to tell the time by using a pole or rod (the gnomon) to cast a shadow on a scale

Telescope The main instrument used to see stars and planets

The Moon The Earth's satellite

The Sun The star which is the central body of the Solar System

Year The time taken for a planet to go once round the Sun

Zodiac A band of constellations stretching around the sky and in which the planets are seen

This is what our planet looks like from space, you can see the British Isles

Index

Published by BBC Educational Publishing
BBC White City
201 Wood Lane
London W12 7TS

First published 1995
© Martin Lunn/BBC Enterprises Ltd 1995
The moral right of the author has been asserted

Paperback ISBN: 0 563 39739 X
Hardback ISBN: 0 563 39738 1

Colour reproduction by Daylight Colour Art, Singapore
Cover origination in England by Goodfellow & Egan
Printed and bound by B.P.C Paulton

Illustrations: © Brigid Collins **(pp. 29, 31, 33, 42-43, 44-45)**; © Line & Line **(pp. 5, 6, 9, 22, 26)**; © The Maltings Partnership **(pp. 4, 8, 13 (top), 24-25)**; © Steve Noon **(pp. 11, 13 (bottom), 14, 32)**
Photo credits: Bridgeman Art Library/Giraudon **p. 2 (top)**; Andrew Cottam **p. 12**; Mary Evans Picture Library **p. 27 (top)**; Luke Finn/BBC Education **pp. 38/39**; Akira Fujii **p. 27 (bottom)**; The Garden Picture Library **p. 4**; Images Colour Library **p. 3**; Kobal Collection **p. 41 (right)**; NASA **pp. 10 (bottom), 11, 17 (bottom), 19 (top), 36 (left), 37 (bottom)**; Science Photo Library **pp. 2 (bottom), 5, 7 (bottom), 9, 10 (top), 14, 15 (top), 16, 17 (top and middle), 19 (middle and bottom), 20, 21, 23, 33, 34, 35, 36 (right), 37 (top), 40, 41 (left), 47**; Roger W Sinnott **p. 15 (bottom)**; Staatliche Kunstsammlungen, Dresden **p. 7 (top)** detail from *The Kingdom of Flora* by Nicholas Poussin.
Front cover: Science Photo Library **(main picture)** *Apollo 11 view of the earth rising over the moon,* **(right)** *launch of space shuttle Atlantis.*

Mylar is a registered trademark of E. I. du Pont de Nemours and Company
Teflon is a registered trade mark of E. I. du Pont de Nemours and Company
WD–40 is a registered trade mark of the WD–40 Company